Twelve-Fatality Nursing Home Fire
Norfolk, Virginia

Investigated by: Randolph E. Kirby
Hollis Stambaugh

This is Report 034 of the Major Fires Investigation Project conducted by TriData Corporation under contract EMW-88-C-2649 to the United States Fire Administration, Federal Emergency Management Agency.

FEMA

Department of Homeland Security
United States Fire Administration
National Fire Data Center

U.S. Fire Administration Fire Investigations Program

The U.S. Fire Administration develops reports on selected major fires throughout the country. The fires usually involve multiple deaths or a large loss of property. But the primary criterion for deciding to do a report is whether it will result in significant "lessons learned." In some cases these lessons bring to light new knowledge about fire--the effect of building construction or contents, human behavior in fire, etc. In other cases, the lessons are not new but are serious enough to highlight once again, with yet another fire tragedy report. In some cases, special reports are developed to discuss events, drills, or new technologies which are of interest to the fire service.

The reports are sent to fire magazines and are distributed at National and Regional fire meetings. The International Association of Fire Chiefs assists the USFA in disseminating the findings throughout the fire service. On a continuing basis the reports are available on request from the USFA; announcements of their availability are published widely in fire journals and newsletters.

This body of work provides detailed information on the nature of the fire problem for policymakers who must decide on allocations of resources between fire and other pressing problems, and within the fire service to improve codes and code enforcement, training, public fire education, building technology, and other related areas.

The Fire Administration, which has no regulatory authority, sends an experienced fire investigator into a community after a major incident only after having conferred with the local fire authorities to insure that the assistance and presence of the USFA would be supportive and would in no way interfere with any review of the incident they are themselves conducting. The intent is not to arrive during the event or even immediately after, but rather after the dust settles, so that a complete and objective review of all the important aspects of the incident can be made. Local authorities review the USFA's report while it is in draft. The USFA investigator or team is available to local authorities should they wish to request technical assistance for their own investigation.

This report and its recommendations were developed by USFA staff and by TriData Corporation, Arlington, Virginia, its staff and consultants, who are under contract to assist the USFA in carrying out the Fire Reports Program.

The USFA appreciates the cooperation received from the Norfolk Fire Department and the Hillhaven Rehabilitation and Convalescent Center. Particular thanks go to Chief Thomas Gardner and Investigator Forest Parham of Norfolk Fire Department and to Assistant Administrator Willie Alston of Hillhaven.

For additional copies of this report write to the U.S. Fire Administration, 16825 South Seton Avenue, Emmitsburg, Maryland 21727. The report is available on the Administration's Web site at http://www.usfa.dhs.gov/

U.S. Fire Administration

Mission Statement

As an entity of the Department of Homeland Security, the mission of the USFA is to reduce life and economic losses due to fire and related emergencies, through leadership, advocacy, coordination, and support. We serve the Nation independently, in coordination with other Federal agencies, and in partnership with fire protection and emergency service communities. With a commitment to excellence, we provide public education, training, technology, and data initiatives.

TABLE OF CONTENTS

Twelve-Fatality Nursing Home Fire
Norfolk, Virginia

Local Contacts: Thomas E. Gardner, Fire Chief
Forest L. Parham, Investigator
Norfolk Fire Department
540 East City Hall Avenue
Norfolk, Virginia 23510
(804) 441-2481

Willie E. Alston, Jr.
Assistant Administrator
Hillhaven Rehabilitation and Convalescent Center
1005 Hampton Boulevard
Norfolk, Virginia 23507
(804) 623-5602

OVERVIEW

On October 5, 1989, at 2218 hours, a fire in Norfolk, Virginia, was reported from the Hillhaven Rehabilitation and Convalescent Home, 1005 Hampton Boulevard. This was a 4-story masonry building, housing 161 elderly patients, most of whom were bedridden.

First arriving firefighting units discovered fire coming out of the window of a second floor patient room located on the front of the building. The fire was lapping up to the third floor window. The second floor was completely filled with heavy smoke, and some flame at the ceiling level was observed. No alarms were heard and there was no apparent commotion.

Second and third alarms were sounded immediately to assist in rescue efforts. Some patients were removed from their rooms by the use of ground ladders set up on the outside. Bedridden patients, trapped in their rooms, had to be carried by firefighters through heavy smoke and heat conditions. Rescue efforts on the second floor required approximately 35 minutes.

SUMMARY OF KEY ISSUES

Issues	Comments
Cause	Believed accidental discarding of lighted match on bed, igniting bed linen and foam rubber pad in second floor patient room.
Detection & Reporting	Fire detected by floor nurse who assisted two occupants from room of origin, leaving door open to hallway, allowing flame and smoke to penetrate hall. Other patients on same floor were made aware of fire by the staff yelling and smoke penetrating their rooms.
Firefighting	Heavy black smoke throughout second floor made search and rescue difficult. Three alarms were needed to provide sufficient staffing.
Building Structure	Sound construction and quick extinguishment prevented structural failure and fire extension to other areas.
Fire Protection Equipment	No smoke detectors in patient rooms. Interior fire alarm system (including detectors) failed to operate, resulting in delayed detection and fire department notification.
	Building equipped with 6-inch standpipe system connected to city water only. No automatic fire pumps. No sprinkler system.
Smoke Barrier Doors	Smoke barrier doors, installed in hallway and equipped with automatic smoke-activated door closures, failed to operate due to blown fuse in fire alarm panel. This allowed rapid spread of heat and smoke throughout second floor.
Code Compliance	Building not subject to current code requirements for fire sprinklers and smoke systems, since it was constructed prior to requirement.
Patient Life Support & Restraint	Many patients restrained to beds with cotton cravats and/or connected to life support systems, making removal by firefighters extremely difficult.
Evacuation	Because of heavy, thick, black smoke and considerable heat, coupled with the fact most occupants were bedridden, rescue and evacuation were difficult and time consuming.
Local Hospital Disaster Plan	Plan was effectively put into operation, thereby providing adequate medical support and transport for the large number of injured.

Approximately 55 patients were removed from the second floor, and eventually, the entire building was evacuated. Heavy smoke conditions claimed the lives of 12 residents and injured 98. In addition, four firefighters were injured.

One hundred thirty-eight fire and rescue services personnel were required to bring the scene under control, officially declared at 0100 hours.

BUILDING STRUCTURE

The building is located in a predominantly residential community in the downtown section of Norfolk, Virginia. It is constructed with brick and cinderblock walls; floors are concrete slabs, supported by steel bar joists. It is a 4-story L-shaped design, 250 feet by 60 feet (see Appendix A for photographs and Appendix B for floor plan).

The building is equipped with service and passenger elevators. The first floor contains the administrative offices, cafeteria, physical therapy treatment rooms, and the building heating and electrical services. The second, third, and fourth floors are devoted to patient rooms, housing 172 beds.

There are three stairwells located on the north, east, and west sides of the building. Each stairwell begins at ground level and terminates at the fourth floor.

Interior décor is largely vinyl-covered and painted wall surfaces, vinyl floor tiles, and a 1-hour rated suspension ceiling.

CODES

The building was constructed in 1969 under the Southern Standard Building Code, which at that time did not require sprinkler systems or smoke detection systems. The building is considered to be in compliance with existing building codes and is not subject to fire-protection upgrading though a fire alarm/smoke detection system was in place. The last inspection by the Norfolk Fire Department was November 1988 at which time reportedly only a few, minor violations were found. The building has enjoyed a very good fire record.

Should this structure be constructed today, complete fire detection and fire sprinkler systems would be required, including smoke detectors in each patient room.

FIRE PROTECTION

The building is equipped with a 6-inch standpipe system located in each exit stairwell. A 2-1/2-inch hose valve is located on each floor level; a 1-1/2-inch valve is located on hallways outside each stairwell.

The building is equipped with an automatic fire alarm system, which is monitored by a private agency. The building has three sets of smoke barrier doors, one set each on floors two, three, and four. These doors are equipped with magnetic hold open devices activated by smoke detectors located in the corridors and interconnected to the fire alarm system.

Exit doors from each floor are equipped with an alarm for the purpose of alerting the nursing staff about wandering patients.

The building does not contain a sprinkler system or individual room smoke detectors. The city water main system in this area is considered to be satisfactory.

ORIGIN AND SPREAD OF FIRE AND SMOKE

The fire originated in Room 226, believed to be as a result of patient accidentally discarding a lighted match onto his bed (after missing the waste can) and igniting the bed linen and the polyurethane mattress pad, which is a highly combustible and smoke-generating material when subjected to open flame. The fire intensified very rapidly, generating tremendous heat and smoke buildup. It was known that the patient was a smoker. The night before the fire he had been caught with cigarettes in his room, which was against the facility's rules.

The room was not equipped with smoke detectors or an automatic fire suppression system, and it appears that the fire burned unabated for a few minutes before it was discovered. A nurse's assistant had checked the patients in Room 226, the room of origin, and then proceeded down the hall to look in on other patients. Originally she stated she was only two rooms away when she smelled smoke and began checking for the source of fire. Later, however, she recalled she was several rooms away from Room 226 and that she checked back into each of these rooms for the fire before finally discovering the blaze in 226.

Once the floor nurse detected the fire she assisted the two occupants from the room. The door remained open, allowing the fire and smoke to penetrate the second floor hall.

Smoke barrier doors, located in the hall and within 20 feet of the room of origin, failed to close, allowing smoke to completely penetrate the second floor. The interior fire alarm system was pulled. Due to a blown fuse in the main fire alarm control panel, that system also failed to operate and no alarm was sounded.

The nurse yelled to other second floor staff that there was a fire. The nurses began to open and close stairwell doors as they attempted to evacuate patients. This allowed smoke to penetrate the upper floors.

It is believed that the fire burned approximately 12-15 minutes before the fire department arrived.

THE FIRE

On October 5, at 2218, a fire call was received by the Norfolk Fire Alarm Dispatch Center from a staff member who worked at the Hillhaven Home. First responding units arrived at the scene in four minutes at 2222. Engine 6's four firefighters went to the front entrance and observed heavy flames from a second story patient room. Engine 6 proceeded with an interior attack from the east side stairwell, advising Engine 7 to make an exterior attack to the room of fire origin.

An immediate call for additional alarms was requested. Engine 7, with four men, arriving moments behind Engine 6, began laying a 5-inch supply line to Engine 6 from a fire hydrant located on Hampton Boulevard at the north end of the building. The fire hydrant was broken and not usable. Engine 1 arrived moments after Engine 7 and proceeded with a 5-inch line to Engine 6. Engine 7 positioned itself at the intersection of Hampton Boulevard and Westover Avenue.

Personnel from Engine 6 carried a highrise pack to the second floor by way of the east stairwell. Hose was connected to the standpipe system and firefighters, who began to make their way to the second floor through the exit stairwell, found the floor completely charged with heavy, black smoke. They observed fire at ceiling level in the area of the smoke barrier doors located midway down the hall. After opening their handline, water was lost for a few moments, probably due to an air pocket within the standpipe system.

By this time, the crew from Engine 7 had entered the building through a window on the second floor and had knocked down the majority of fire. It then became apparent to the members of Engine 6 that the floor had not been evacuated, and that many patients were still in their rooms. Though fire was no longer a threat, dense toxic smoke pervaded the second floor corridor in the location of the room of origin, highly threatening to the frail, elderly residents.

At this time, a nurse was attempting to roll a patient in a bed to the exit stairwell. Firefighters quickly removed the patient from the bed and helped the nurse and patient to the outside. Firefighters began carrying patients, most of whom were bedridden, from smoke-filled rooms. This proved to be a tremendously difficult and time-consuming task, given the smoke conditions and the number of people who needed to be evacuated.

As additional fire department personnel arrived, a command post was established in front of the building on Westover Avenue. Three ground ladders were placed against the front wall of the building, where several patients were removed through second floor windows.

A relay system was utilized to remove people from the second floor. Firefighters wearing breathing apparatus took patients from their rooms to the stairwell, where they were transferred to other personnel who carried them to the outside.

By this time, the medical director for the Paramedic Rescue Squad and a Norfolk General Hospital physician arrived and established a triage site on the lawn near the east end of the building.

Rescue was continuing. Firefighters were experiencing difficulty releasing restrained patients from beds, as the restraining devices had to be cut or untied, requiring additional time. Difficulty was also experienced when removing life support systems and body fluid tubes, which were connected to bed and patient. Because it took so long to remove the bed straps and to disconnect patients from medical equipment, and because rescuers had to move cautiously down the stairwell carrying elderly, infirm patients, a traffic jam developed in the hall outside the stairwell. This further complicated rescue operations.

As the fire suppression and evacuation effort on the second floor was proceeding, fire personnel were stationed on the third and fourth floors, where moderate smoke had permeated. They, along with nursing staff, began reassuring patients and closing doors to rooms. The fire department then hooked up their new high-volume smoke removal unit (Air-1) to the front entrance, and used the unit to blast smoke out of the building.

By approximately 2240, the fire on the second floor was extinguished and most patients there had been removed. The medical director ordered evacuation of the third and fourth floors as a precaution since some patients were showing signs of distress.

Thirty-four ambulances, from Norfolk, Chesapeake, Portsmouth, and local Navy bases were used to transport the injured to area hospitals. Two Navy ambulance buses and local transit minivans also were made available to transport wheelchair patients. The hospital emergency procedure that was put into operation apparently worked quite effectively, as medical treatment for so many was given without delay. Nevertheless, some patients were pronounced dead at the triage site; others at hospitals later.

The building was occupied by 161 patients and 28 staff members on the evening of the fire. This emergency required the services of approximately 138 fire and rescue service personnel. The scene was officially declared under control at 0100 hours.

FATALITIES

Twelve elderly patients, most of whom were bedridden, died as a result of smoke inhalation or other complications directly related to exposure from heat and smoke. The nine women, ranging in age from 71 to 97 years, and three men (including the patient who started the fire) ranging in age from 65 to 92 years, died either at the scene or in the hospital sometime later. All the victims resided on the second floor in the immediate vicinity of the fire origin. Of the original seven fatalities at the scene, all were reported to have carboxyhemoglobin rates of 54-59 percent, according to doctors at the hospital.

INJURIES

Building Occupants -- Ninety-eight patients were injured and required hospital treatment. Three died later within days of the incident. Most injuries were due to smoke inhalation problems. Four were considered critical. The majority of those injured resided on the second floor. Others lived on the third and fourth floors, where there was some penetration of smoke.

Firefighter Injuries -- There were four firefighters injured, all as a result of rescue efforts. Three were treated for smoke inhalation, and one was treated for a strained back. All were treated at the hospital and have since returned to duty.

DAMAGE ASSESSMENT

Dollar loss is estimated at $100,000. Fire completely gutted the room of origin and caused moderate fire damage to ceiling and walls on portions of the second floor. There was heavy smoke damage throughout the second floor, with moderate smoke damage to the third and fourth floors. There was no apparent structural damage.

Additional damage was prevented by rapid extinguishment by the fire department, coupled with the practice of closing patient room doors by the nursing staff and sound construction of the building.

LESSONS LEARNED

1. **Institutional buildings, regardless of when they were built, need full built-in protection.**

    Regardless of when they were constructed, multiple occupancy institutional buildings should be subjected to current fire codes regarding installation of fire protection equipment. This fire is further testimony to the urgent need for such action. The installation of a fire sprinkler system, coupled with a well-designed smoke detection system, would have reduced, if not eliminated, this tragic loss of life.

    When dealing with large numbers of frail and bedridden people in an institutional setting, evacuation often is not a viable alternative. As such, there is an essential need for facilities such as Hillhaven to have complete automatic fire suppression and detection capabilities.

2. **Frequent testing of fire protection and alarm systems is critical.**

    The fact that the smoke barrier doors and alarm system failed to operate illustrates the importance of frequent and thorough inspections and testing of fire protection systems.

    Nursing home operators should be made aware of the importance of making sure the safety systems are operating, and of their self-interest in preventing damage and liability suits.

3. **Flammable furnishings contribute to rapid fire growth and flashover.**

    While the polyurethane mattress pads apparently were treated for fire retardancy, they were a major factor in heat and smoke buildup. Equipment and decorations, such as drapes, wall coverings, bed linen, etc., should be of the type that affords the lowest flame spread possible.

4. **Commonly used patient restraints seriously hamper evacuation efforts during emergencies.**

    Restraining patients to beds should be accomplished by using a type of restraint that can be released with relative ease and speed in the event of emergencies. The use of cotton cravats in this fire hampered firefighters in their rescue efforts, as straps had to be cut or untied before patients could be evacuated.

Consideration also must be given to the method used in attaching life support systems to patients and beds.

5. It is important to remember to rotate personnel at the scene.

Personnel from the first arriving engine company were also among the last to leave and were quite exhausted. This is a reminder of the need to rotate personnel as feasible, so as to avoid overexertion and potential injury.

6. Employee training and practice drills pay off when an emergency does happen.

The Hillhaven fire once again demonstrates the importance of developing and implementing a well-designed emergency procedures program. The program at this facility was excellent. It was well-designed, clearly documented, and was practiced on a monthly basis. This training was evident the night of the fire when staff immediately began closing doors to impede the spread of fire and smoke and attended to patients removed to the lawn. Their efforts went a long way toward effecting a prompt and efficient response to the fire and in limiting confusion at the scene.

APPENDICES

A. List of Slides, Selected Photographs

B. Floor Plan of Second Floor (Layout of third and fourth floors is similar.)

C. Fire Scene Diagram Showing Fire Units' Positions at Fire

D. Unit Response Times

E. "Egg Crate" Pad Label

F. Units Used at the Fire

G. Fire Incident Report

H. Fire Department Pre-fire Plan Floor Diagram for Hillhaven

I. Sample of Hillhaven's Safety Committee Meeting Minutes, Monthly Training Schedule, and Staff Attendance Record

APPENDIX A

List of Slides, Selected Photographs

Slides and photographs are included with the master report at the USFA. Below the slides with an asterisk have been made into photos and are presented following this list.

*1. Main entrance to Hillhaven Home.

2. South portion of building, facing Westover Avenue.

*3. South end of building, indicating second floor window where fire originated.

4. East end of building facing Hampton Avenue.

*5. Northeast end of building looking southwest (note defective fire hydrant).

6. West end of building and parking lot.

7. Typical standpipe riser for building.

*8. Typical 1-1/2-inch hose outlet, located at each stairwell entrance on each floor.

*9. Main control panel for fire alarm and smoke detector systems.

10. Emergency generator set.

11. Siamese connection to standpipe system on east side of building (note obstructions).

*12. Typical door alarm on each stairwell door.

13. Typical bed used throughout home.

*14. "Egg crate" polyurethane mattress used throughout home.

15. Hallway looking north from front of building on third floor.

16. Hallway looking north from front of building on third floor.

*17. Smoke detector and smoke barrier located on second, third, and fourth floors.

*18. Fire damage to smoke barrier doors and to ceiling from room of origin on right.

*19. Fire and heat damage to ceiling and walls on opposite side of smoke barrier doors.

20. Magnetic hold open device for second floor smoke barrier doors.

21. Fire damage to hallway from room of origin.

22. Point of origin in Room 226.

23. Fire damage to front wall of Room 226.

24. Fire damage to wall in Room 226.

25. Fire damage to ceiling system in Room 226 (note relatively good condition of steel bar joists).

26. Fire damage in Room 226 facing hallway.

1. Main entrance to Hillhaven Home.

3. South end of building, indicating second floor window where fire originated.

5. Northeast end of building looking southwest (note defective fire hydrant).

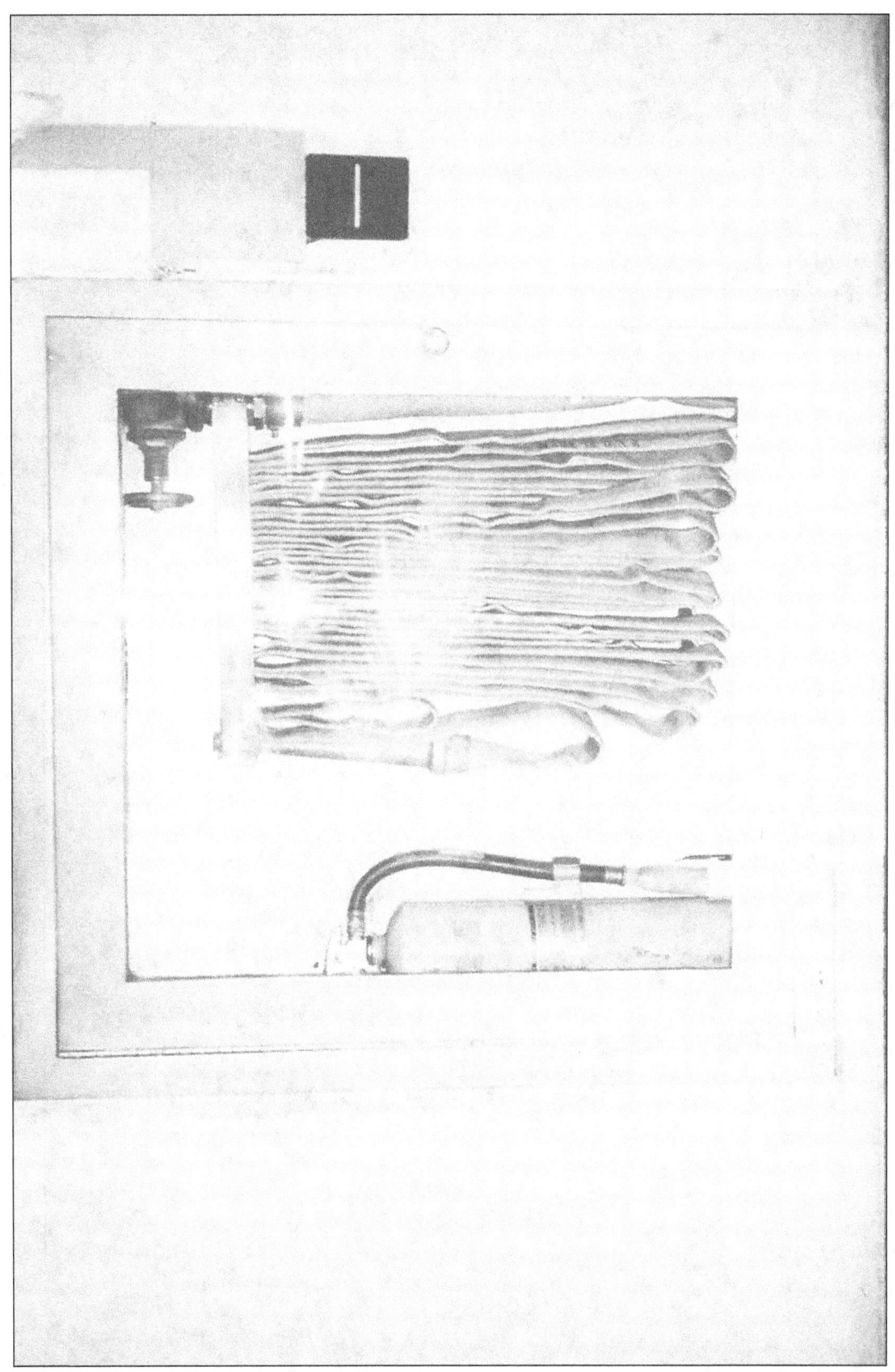

8. Typical 1-1/2-inch hose outlet, located at each stairwell entrance on each floor.

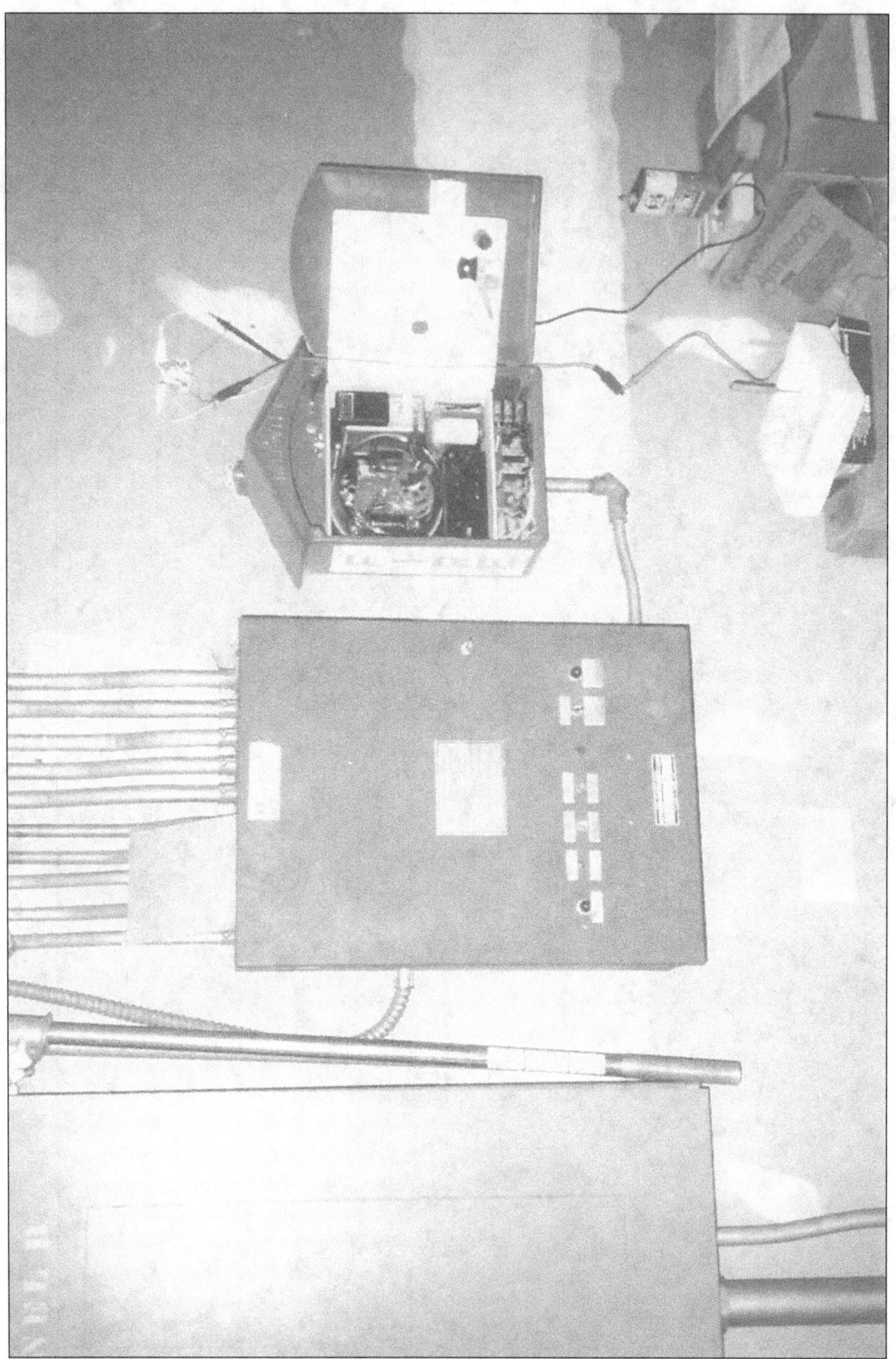

9. Main control panel for fire alarm and smoke detector systems.

12. Typical door alarm on each stairwell door.

14. "Egg crate" polyurethane mattress used throughout home.

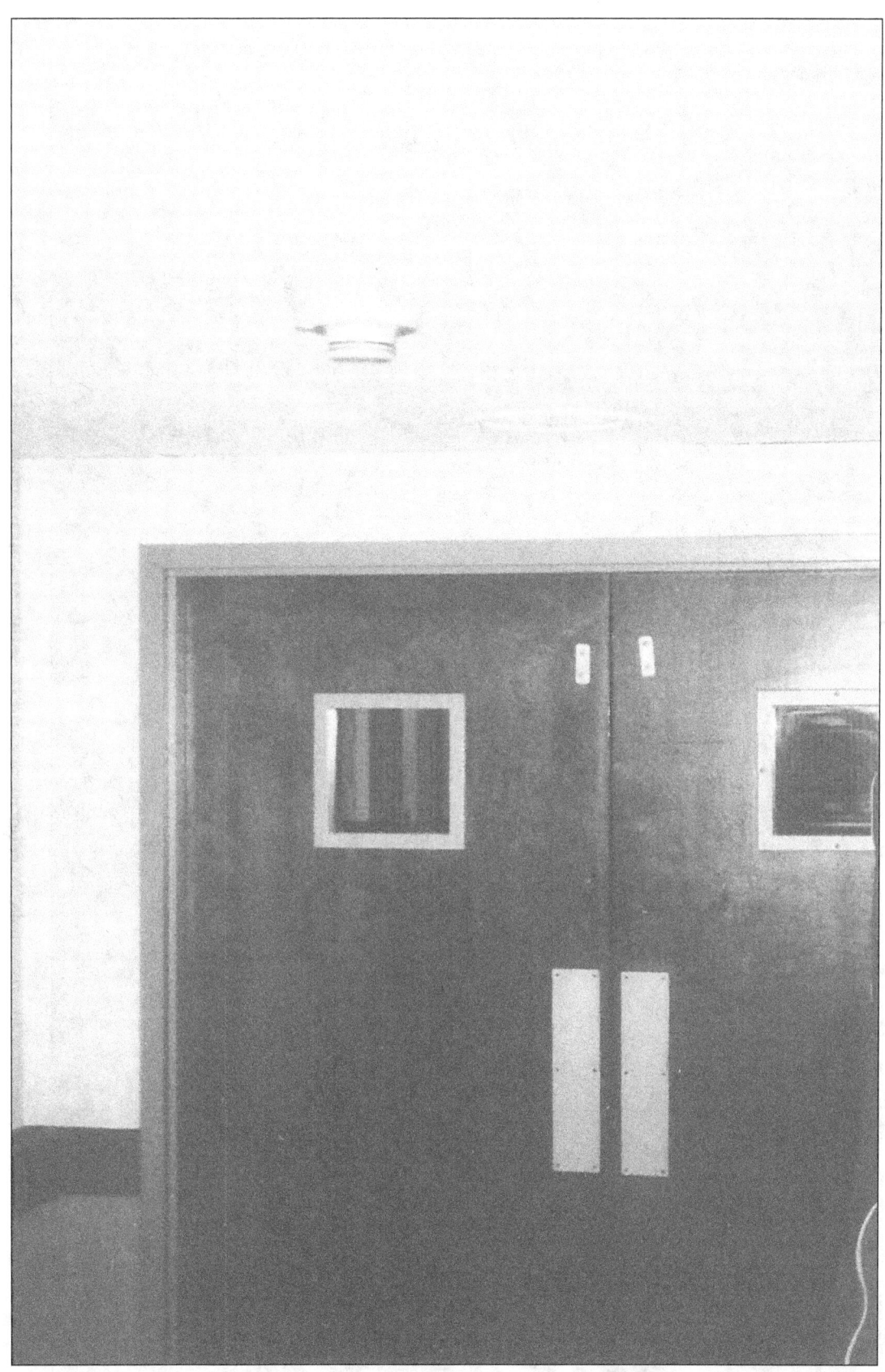

17. Smoke detector and smoke barrier located on second, third, and fourth floors.

18. Fire damage to smoke barrier doors and to ceiling from room of origin on right.

19. Fire and heat damage to ceiling and walls on opposite side of smoke barrier doors.

APPENDIX B

Floor Plan of Second Floor

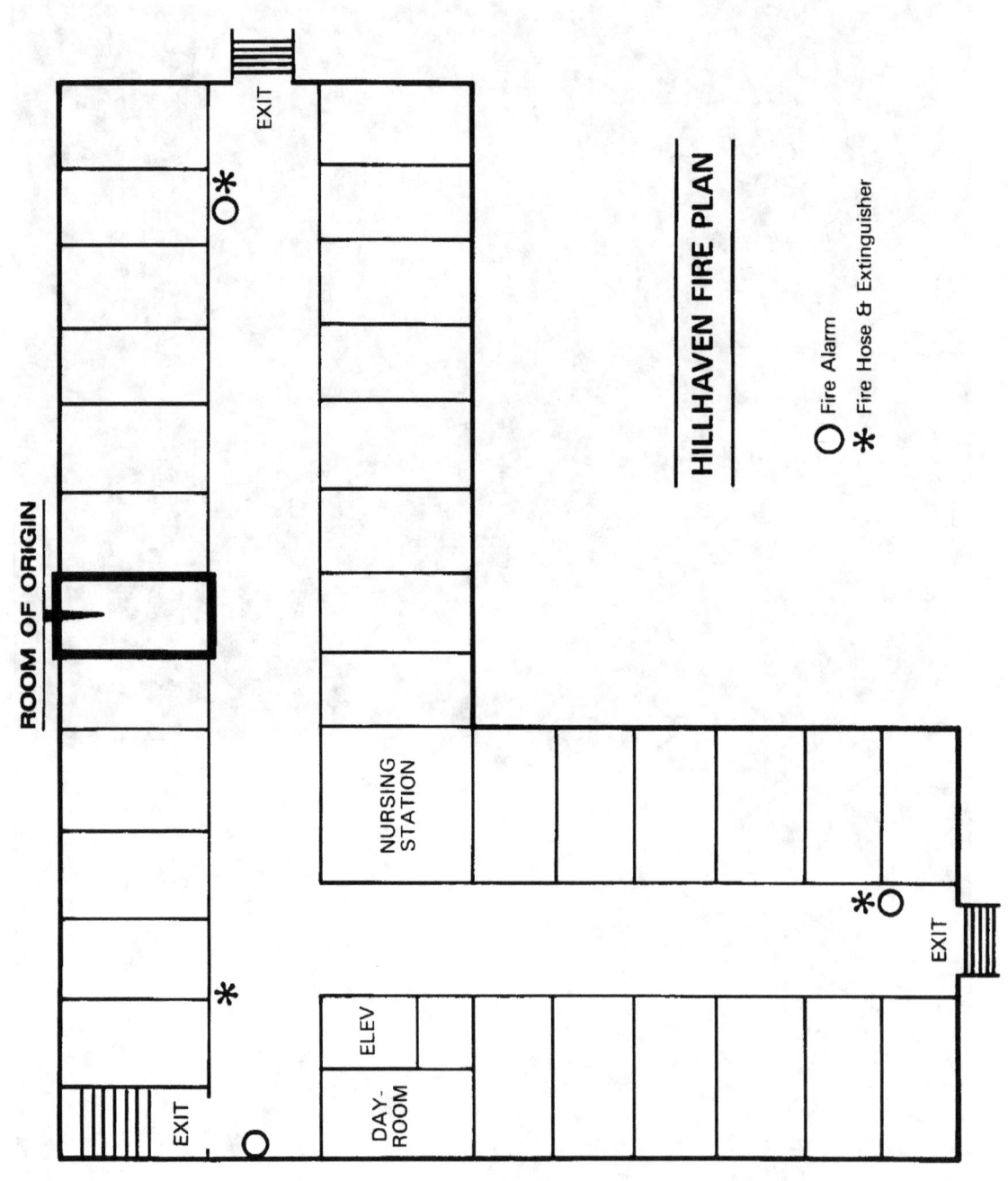

HILLHAVEN FIRE PLAN

○ Fire Alarm
* Fire Hose & Extinguisher

ROOM OF ORIGIN

EXIT

NURSING STATION

ELEV

DAY-ROOM

EXIT

EXIT

Fire Scene Diagram Showing Fire Units' Positions at Fire

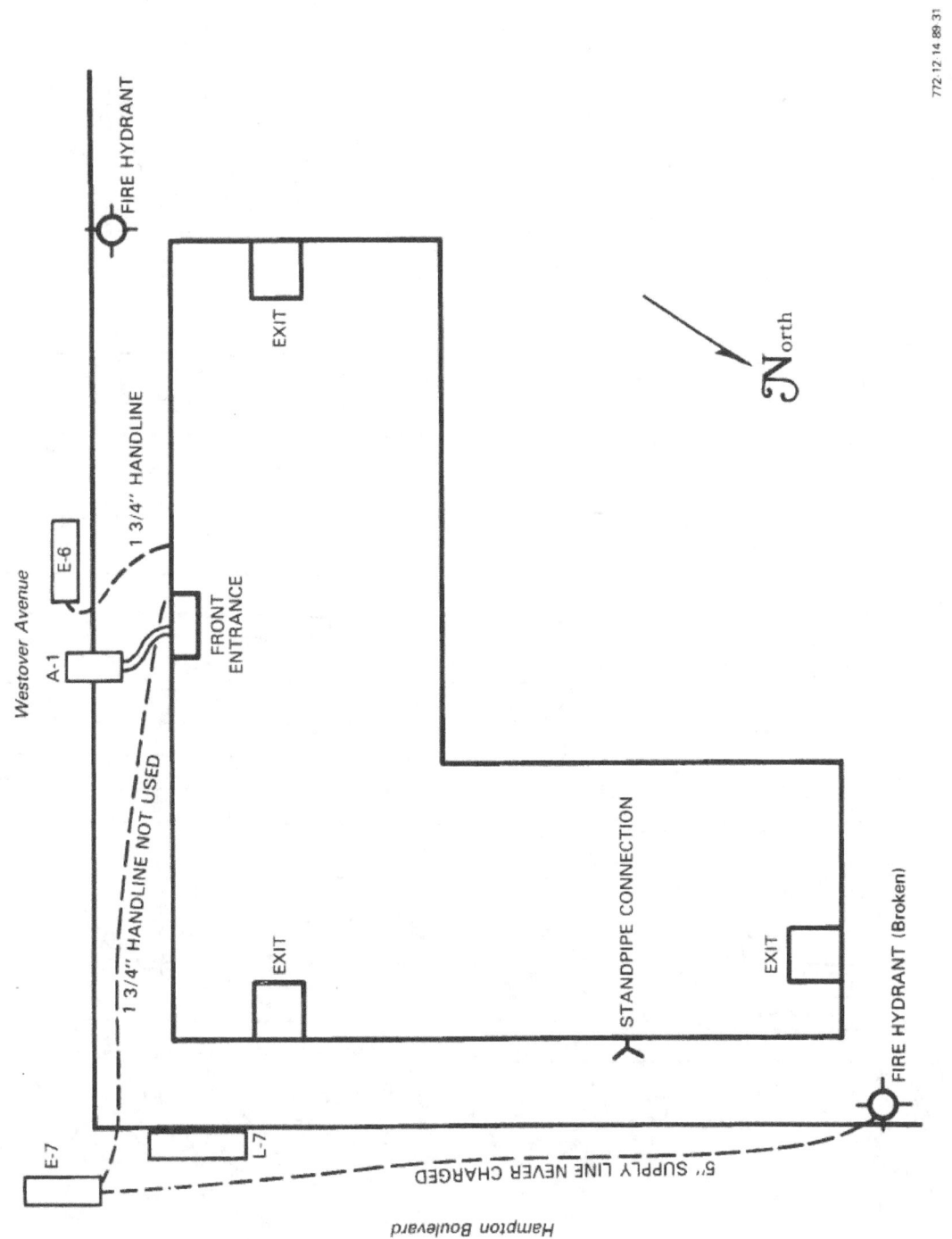

OES-5 **EMERGENCY SERVICES CCC**
FIRE ALARM REPORT

page #1 — № 11052
Incident No. ....

Location		Apt. No.
Hillhaven 1005 Hampton		

Source of Alarm		Type
1 2 3 4 5 6 (7)		10– /

Rec'd	Chan. 1	Chan. 2	Chan. 3	
70	31	31	31	450

			Received
		2218	Broadcast
		2219	1st 10-9
		2222	10-8
		0433	

Chief: S. Gray

Units No.	Response	10–9	10–8
E C 6 # /		2222	*0433
E C 7 /		2223	0055
E L 7 /		2223	0056
S C /		?	0415
C 0 /		2225	0412
E 0 2 2227		?	0409
E 0 1 2227		?	0356
E 0 8 2227		?	0125
L 0 2 2227		?	0103
L 0 / 2227		2234	0414
A 4 / 2226		2238	0059
C L 0 3 2227		2234	0535

Remarks:
3 Alarm *4th alarm
per
R. L. Smith

EMERGENCY SERVICES CCC
FIRE ALARM REPORT

page #2 — № 11052
Incident No. ....

Location		Apt. No.
1005 Hampton Blvd		

Source of Alarm		Type
1 2 3 4 5 6 (7)		10– 1

Rec'd	Chan. 1	Chan. 2	Chan. 3	
70	143	143	143	450

			Received
			Broadcast
			1st 10-9
			10-8

Chief: SYRAX

Units No.	Response	10–9	10–8
E 1 1	2234	*2242	0037
E 1 2	2234	?	0012
E 0 9	2240	?	0009
E 1 0	2240	?	0038
E 1 4	2249	?	0002
E 1 5	2249	?	0043
S 0 2	2249	?	0215
8 6 0	2251	2302	0405
8 6 3	2251	2329	0246
C 100	2250	2309	0257
F 1 0	Responded		
C 3 1	Responded		

Remarks:
NUB - standby Sta12
Control (0010)*

"Egg Crate" Pad Label

EGGCRATE® PAD

EGGCRATE® IS A REGISTERED TRADEMARK OF BIO CLINIC

14060

013350   4/00

CARE AND USE INSTRUCTIONS

- THIS PAD SHOULD NOT BE LAUNDERED.
- LAUNDERING REDUCES FIRE RETARDANCY OF THE PAD.
- DO NOT SMOKE WHILE IN BED OR ON THIS PAD.
- THIS PAD IS INTENDED FOR "SINGLE PATIENT USE ONLY."

ALL NEW POLYURETHANE - CA 28879

COMBUSTION MODIFIED FOR CALIF., TECHNICAL BULLETIN #117

10670 Acacia
Rancho Cucamonga,
CA 91730

BIO CLINIC
SUNRISE MEDICAL

800-854-2369
714-969-2535
Label #2242

APPENDIX F

Units Used at the Fire

11	Engines
3	Ladder Trucks
2	Squad Rescue Units
39	Ambulances
2	Navy Ambulance Buses
1	Safety Officer
6	Minivans with Chair Lifts
1	High Volume Ventilation Unit
138	Personnel (approximate)

APPENDIX G

Fire Incident Report

PLEASE PRINT IN YOUR OWN WORDS BOTH A
WRITTEN AND CODED RESPONSE WHEN APPLICABLE.
MARK BOXES WITH AN "X" AND CODE CHOICES.
LEAVE NO ITEMS BLANK.

Appendix G
FIRE INCIDENT REPORT

NORFOLK _____ Fire Department

1 ☐ DELETE
2 ☐ CHANGE

A | 10 | FDID 7 0 0 6 | INCIDENT NO. 0 1 1 0 5 2 | EXP. NO. 0 0 | MO. 1 0 | DAY 0 5 | YR. 8 5 | DAY OF WEEK 7☐S 1☐S 3☐T 5☐T 2☐M 4☐W 6☐F | ALARM TIME 2 2 : 1 8 | ARRIVAL TIME 2 2 : 2 2 | TIME-IN SERVICE 0 4 : 3 4

B TYPE OF SITUATION FOUND (pg. 17)
11 ☒ structure fire
12 ☐ outside of structure fire
13 ☐ vehicle fire
14 ☐ tree, brush, grass fire
15 ☐ refuse fire
16 ☐ explosion no after-fire
17 ☐ outside spill with fire
21 ☐ steam rupture
22 ☐ air, gas rupture
31 ☐ inhalator call
32 ☐ emergency medical call
33 ☐ lock-in
34 ☐ search
35 ☐ extrication
41 ☐ spill, leak - no fire
44 ☐ power line down
45 ☐ arcing, shorted elec. equip.
46 ☐ aircraft standby
47 ☐ chemical emergency
51 ☐ lock-out
52 ☐ water evacuation
53 ☐ smoke removal
54 ☐ animal rescue
55 ☐ assist police
56 ☐ unauthorized burning
57 ☐ move-up, cover assignment
61 ☐ smoke scare
62 ☐ wrong location
63 ☐ controlled burning
64 ☐ vicinity alarm
65 ☐ steam, gas mistaken for smoke
71 ☐ malicious false
72 ☐ bomb scare
73 ☐ system malfunction
74 ☐ unintentional
☐ other _____

TYPE OF ACTION TAKEN
1 ☒ extinguishment
2 ☐ rescue
3 ☐ investigation only
7 ☐ remove hazard
6 ☐ salvage
8 ☐ fill in, move up, transfer
5 ☐ not classified above
0 ☐ undetermined or not reported

MUTUAL AID
1 ☐ received
2 ☐ given
none (no code is necessary)

C FIXED PROPERTY USE (pg. 22)
151 ☐ restaurant
411 ☐ 1 family, year-round
414 ☐ 2-family, year-round
422 ☐ apartment, 3-6 units
423 ☐ apartment, 7-20 units
424 ☐ apartment, over 20 units
655 ☐ crops, orchards
661 ☐ forest/timber w/o logging operation
915 ☐ vacant property
931 ☐ open land, field
932 ☐ dump/sanitary landfill
936 ☐ vacant lot
951 ☐ railroad right of way
961 ☐ highway, limited access
962 ☐ paved public street
963 ☐ paved private street
984 ☐ unpaved street/road/path
985 ☐ uncovered parking area
31L ☐ other _____
CARE OF AGED W/NURSIN 3/2FT.

IGNITION FACTOR (pg. 44)
11 ☐ incendiary/not during civil disturbance
21 ☐ suspicious/not during civil disturbance
31 ☒ abandoned materials
34 ☐ uncontrolled open fire
36 ☐ children playing w/heat of ignition
46 ☐ combustible too close
51 ☐ part failure
54 ☐ short circuit
55 ☐ other electrical failure
56 ☐ lack of maintenance
57 ☐ backfire
95 ☐ property too close
73 ☐ unattended
92 ☐ rekindled
00 ☐ undetermined
☐ other ____
FOR NON-FIRES ONLY (No Code)

3 1 1

3 1

COMPLETE ON ALL INCIDENTS

D CORRECT ADDRESS (Up to Maximum of 21 Characters)
1 0 0 5 H A M P T O N B L V D
ZIP CODE 2 3 5 0 7
CENSUS TRACT

E | 11 | OCCUPANT LAST NAME _HILLHAVEN NURSING HOME-Vickie_ FIRST _ARCHER_ MI | TELEPHONE 623-5602 | ROOM OR APT. N/A

F | 12 | OWNER LAST NAME _HILLHAVEN CORP._ FIRST MI | ADDRESS 1023 LASKIN RD. VA B. | TELEPHONE 425-1311

G | 13 | METHOD OF ALARM FROM PUBLIC
1 ☐ telephone direct
2 ☐ municipal alarm system
3 ☐ private alarm system
4 ☐ radio
5 ☐ verbal
6 ☐ no alarm rec'd
7 ☒ tie-line (911)
8 ☐ voice signal municipal alarm signal
9 ☐ not classified above
0 ☐ undetermined or not reported
17 | CO-INSPECTION DISTRICT 950 | SHIFT C | NO. ALARMS 4

H NO. FIRE SERVICE PERSONNEL RESPONDED 074 | NO. ENGINES RESPONDED 011 | NO. AERIAL APPARATUS RESPONDED 003 | NO. OTHER VEHICLES RESPONDED SEE BACK REMARKS 011

COMPLETE IF CASUALTY OR FIRE

I | 20 | NO. INCIDENT-RELATED INJURIES (COMPLETE VFIRS 3) FIRE SERVICE 003 (COMPLETE VFIRS 2) OTHERS 143 | NO. INCIDENT-RELATED FATALITIES (COMPLETE VFIRS 3) FIRE SERVICE 000 (COMPLETE VFIRS 2) OTHERS 009

J COMPLEX (pg. 61)
41 ☐ dwelling
42 ☐ apartment
58 ☐ shopping
65 ☐ farm
98 ☐ road
99 ☐ no complex
32 ☒ other 1 3
MOBILE PROPERTY TYPE (pg. 63) (COMPLETE LINE S)
08 ☒ not applicable
11 ☐ automobile
17 ☐ mobile home
22 ☐ pick-up truck
☐ other 0 8

COMPLETE FOR ALL FIRES

K ARSON AREA OF FIRE ORIGIN (pg. 86)
14 ☐ lounge area
21 ☒ sleeping area/under 5 persons
24 ☐ kitchen/cooking area
46 ☐ trash area/container
57 ☐ chimney
81 ☐ vehicle passenger area
83 ☐ engine/running gear area
92 ☐ on/near street, public way
94 ☐ lawn/field/open area
00 ☐ undetermined
☐ other 2 1
EQUIPMENT INVOLVED IN IGNITION (COMPLETE LINE T) (pg. 70)
13 ☐ heating stove
18 ☐ chimney, gas vent flue
21 ☐ cooking stove
96 ☐ vehicle
☒ N/A
00 ☐ undetermined
☐ other 9 8

L FORM OF HEAT OF IGNITION (pg. 74) (HEAT SOURCE)
11 ☐ heat from solid fuel
24 ☐ short circuit
31 ☐ cigarette
45 ☒ match
48 ☐ backfire from engine
4 1 5
TYPE OF MATERIAL IGNITED (pg. 77) (COMPOSITION)
34 ☐ creosote
54 ☐ grass, leaves
63 ☐ sawn wood
67 ☐ paper
00 ☐ unknown
72 ☒ 1 7 2
FORM OF MATERIAL IGNITED (pg. 80) (USE)
21 ☐ upholstered sofa, chair
65 ☐ fuel
74 ☐ waste, creosote
32 ☒ other BEDDING 3 9

M METHOD OF EXTINGUISHMENT
1 ☐ self-extinguished
2 ☐ make-shift aids
3 ☐ portable extinguisher
4 ☐ automatic ext. system
5 ☐ pre-connect hose/tank
6 ☐ pre-connect hose/hydrant
7 ☒ hand-laid hose/hydrant
8 ☐ master stream device
9 ☐ not classified above
0 ☐ undetermined
7
LEVEL OF FIRE ORIGIN
1 ☐ grade level to 9 ft.
2 ☒ 10 to 19 feet
3 ☐ 20 to 29 feet
4 ☐ 30 to 49 feet
5 ☐ 50 to 70 feet
6 ☐ over 70 feet
7 ☐ objects in flight
8 ☐ below ground level
9 ☐ not classified above
0 ☐ undetermined
ESTIMATED TOTAL DOLLAR LOSS (COMPLETE LINE V)
2 0 0 0 1 0 0 0 0 0 0

N NUMBER OF STORIES
1 ☐ 1 story
2 ☐ 2 stories
3 ☒ 3 to 4 stories
4 ☐ 5 to 8 stories
5 ☐ 7 to 12 stories
6 ☐ 13 to 24 stories
7 ☐ 25 to 49 stories
8 ☐ 50 stories or more
0 ☐ undetermined
3
CONSTRUCTION TYPE
1 ☐ fire resistive
2 ☐ heavy timber
3 ☒ protected non-combustible
4 ☐ unprotected non-combustible
5 ☐ protected ordinary
6 ☐ unprotected ordinary
7 ☐ protected wood frame
8 ☐ unprotected wood frame
9 ☐ not classified above
0 ☐ undetermined
3

COMPLETE IF STRUCTURE FIRE

O EXTENT OF DAMAGE | Flame | Smoke
confined to object of origin 1☐ 1☐
confined to area of origin 2☐ 2☐
confined to room of origin 3☐ 3☐
confined to fire-rated comp. 4☐ 4☐
confined to floor of origin 5☒ 5☐
confined to structure of origin 6☐ 6☒
extended beyond structure 7☐ 7☐
undetermined or not reported 0☐ 0☐
no damage of this type (N/A) 9☐
Flame 5 | Smoke 6

P DETECTOR PERFORMANCE
1 ☐ det. in room or space of fire origin -oper.
2 ☐ det. not in room or space of fire origin -oper.
3 ☒ det. in room or space of origin -no oper.
4 ☐ det. not in room or space of origin - no oper.
5 ☐ det. in room or space of fire origin but fire too small to oper.
6 ☐ no detectors present (N/A)
9 ☐ not classified above
0 ☐ undetermined
3
SPRINKLER PERFORMANCE
1 ☐ equipment operated
2 ☐ equipment should have operated -did not
3 ☐ equipment pres. but fire too small to oper.
8 ☒ no equipment present (N/A)
9 ☐ not classified above
0 ☐ undetermined
9

Q SEO PARHAM REDFIELD
TYPE OF MATERIAL GENERATING MOST SMOKE (pg. 102)
63 ☐ sawn wood
72 ☒ cotton/rayon
98 ☐ N/A
00 ☐ unknown
☐ other 7 2
AVENUE OF SMOKE TRAVEL
1 ☐ air handling duct
2 ☒ corridor
3 ☐ elevator shaft
4 ☐ stairwell
5 ☐ opening in construction
6 ☐ utility opening in wall
7 ☐ utility opening in floor
8 ☐ no avenue of smoke travel (N/A)
9 ☐ not classified above
0 ☐ undetermined or not reported
2
FORM OF MATERIAL GENERATING MOST SMOKE
21 ☐ upholstered chair, sofa
44 ☐ papers, magazines
61 ☐ electrical wire insulation
75 ☐ waste, creosote
98 ☐ N/A
32 ☒ other 3 2

S | 30 | IF MOBILE PROPERTY N/A | YEAR | MAKE | MODEL | SERIAL NO. | LICENSE NO. (IF ANY)

T | 40 | IF EQUIPMENT INVOLVED IN IGNITION N/A | YEAR | MAKE | MODEL | SERIAL NO.

V | 60 | ESTIMATED PROPERTY VALUE 0 0 1 0 0 0 0 0 0 | Officer in Charge (Name, Position, Assignment) _BLACKSTONE N.C. Act. ASS'T CH+R_ 103 Date 10-5-85
Member Making Report (If Different from Above) Date

White Copy to the Department of Fire
☐ Check box if Remarks

ALL INCIDENT

25

Fire Department Pre-fire Plan Floor Diagram
for Hillhaven

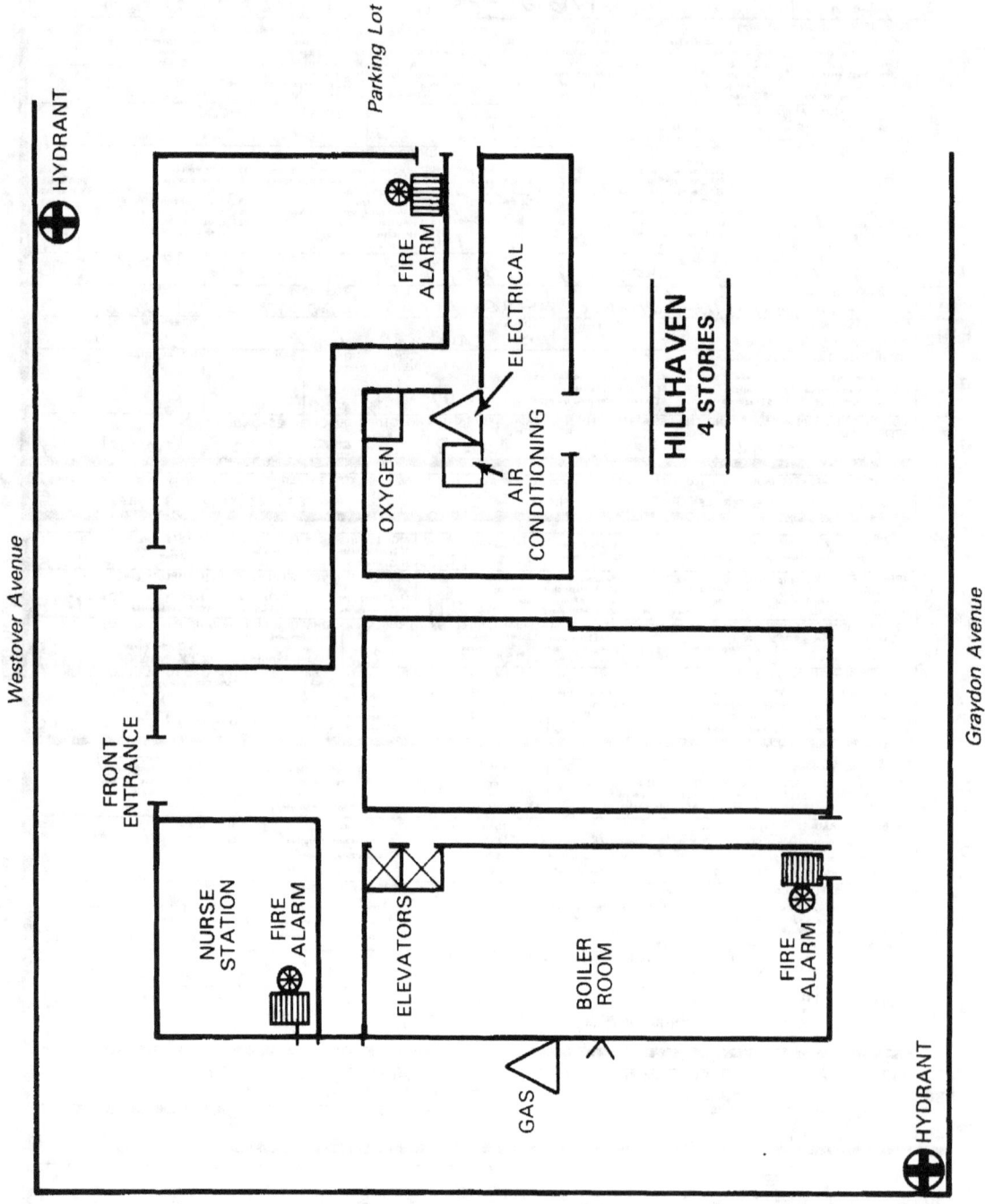

APPENDIX I

Sample of Hillhaven's Safety Committee Meeting Minutes and Staff Attendance Record

SAFETY COMMITTEE MEETING MINUTES

FACILITY _HRCC 826_ CITY _Norfolk, Va._ DATE _9/7/89_

CHAIRPERSON _Susan Foster_

1. Members:
   Althea M. Hyman _____ _____ Willie Winston
   Linda White _____ _____ Rufus Farvas
   Franklin Zulichi _____ _____
   _____ R. Phillips RNC DNS
   _____ Ackie Archer

2. Visitors:

3. Read minutes of last meeting and correct if necessary.
4. Unfinished business (status of previous recommendations, programs, etc.).
5. Review of incidents (Patient and Employee) with recommendations.
6. Inspections and subsequent recommendations.
7. New Business.
8. Safety Education and Motivation.

Old Business: None to Report

New Business ① Fire Drill was again held in Rehab Dept Sept on 7-3 The response from Rehab Dept was greatly improved. Some staff were on the 1st floor & the time & were shown the picture to be told which indicated a N/A & nursing staff member were to report to their unit during a fire. More education will done on this for all staff members ② Review of Employee incidents report attached ③ Review of pt. statistics - report attached ④ Environmental check list was completed & presented by Blaustein Env Serv Service, noted was Activities fire Extinguisher is in area where it can be concealed & a chair if there is not properly positioned. some shower tiles, tiles in Rm 236, 224 were loose. Maintenance aware. Report attached. ⑤ This being a Hurricane season again the ER Preparedness information is readily available in see is all dept books. The Norfolk Shelter & Tracking map is on the Employee's Bulletin Board & some were given out @ this meeting Now available in see office. ⑥ Discussion re the OSHA inspection was had & information that the inspector plans to return in 30 day
(Cont.

Send Copy of Minutes to District Director and Regional Safety Coordinator.

H-3000 (7-88)

27

APPENDIX I (cont'd)

(Cont) Safety #26 HRCC   page #2.

to re-evaluate Maintenance Dept. all dept are again reminded to have MSDS forms available, all chemicals logged ~~current~~ updated all staff in their departments to ~~for~~ have current information pertaining to Universal Precautions & Safety. Complete information re the results of the inspection done Aug 28 & 29 by OSHA to be given next meeting.

② The warm fuzzy program identified 2 more residents needing to be added to list & 1 resident removed @ this time @ wandering resident program where pictures are taken of residents & placed in charts identified 3 new residents added this month.

APPENDIX I (cont'd)